How to Master the Art of Beekeeping!

How to Master the Art of Beekeeping!

How To...
MANIFEST A MIRACLE

Manifesting Reality Isn't Hard Work After All.
Discover My 100% Guaranteed Step-By-Step "Magic"
Formula To Manifest More Money, Love & Good
Health Than You Ever Dreamed Possible!

GARY EVANS

How To...
MANIFEST A MIRACLE

TRUE SECRETS OF
THE MANIFESTATION
PROCESS AND HOW YOU
CAN PUT IT TO WORK
FOR YOU

DOWNLOAD NOW

How To...
MANIFEST A MIRACLE

How to Master the Art of Beekeeping!

Contents

Family Owned Beekeeping Companies

Beekeeping supplies

Companies that have beekeeping stuff deal with all the equipment that is required for this business, like attire for bee keeping which is essential from head to torso, full body suits and just head gear. Along with this equipment they also sell journals and books on beekeeping to help people to understand this field better. Some of the better known beekeeping companies have been in the business for more than a hundred years.

Books for providing short courses in beekeeping

Beekeeping families have a lot of subsidiaries to the beekeeping business too. They use the bees wax to make candles and collect honey which is sold locally in the stores. The books have information on candle making and collecting honey too apart from instructions on the main business which is bee keeping. These family businesses are mostly long standing ones and continue from one generation to the other.

Businesses have advanced technologically

Stepping into the shoes of their ancestors has not been easy for many of the families. Only those families where the parents thought of the new technologies and got their business mechanized with the latest equipment, and advertised with on the internet with their own websites so that they could bring in more customers and higher profits, was the business profitable and worth running. For others the cost of maintaining an old business was too high and they could not handle the expenses anymore. As most descendants of beekeeping families knew no other business they preferred to stick to it, and if the financial situation was not good, then they became commercially owned businesses

A business that once started as a hobby, with just about enough honey to put on toast and biscuits in place of marmalade, became a big time business with the popularity of the use of honey. Earlier honey was not a very popular edible thing as people used sugar and molasses for sweetening their food, but once they found that honey was much cheaper, beekeeping became popular and was taken up by many as an agricultural business. However, now the only problem is that people worry about the safety of the honey that is collected because of all the pesticides that are used. Now with organic foods being more popular and a movement by organic food growers, there is a lot of honey that is free of pesticides. Now many families owned companies supply bees wax for candle making and cosmetics to other manufacturers of these commodities.

California's Almond Orchards

Perfect almond growing conditions in California

Almonds were initially found in central Asia and China, far away from California. Almonds were first brought to California by Franciscan priests, but they were not very successful in cultivating almonds here. It was almost 200 years later in 1900, that those who were interested in almonds found that the climate and weather conditions in California were perfect for almond cultivation. Almost 6 thousand almond growers in the state, cultivated over half a million acres of land to cultivating almond orchards that were genetically far superior to what was available earlier. Almonds are grown successfully for commercial use only in California, and the main reason is the climate in this state which is most suitable for growing almonds. Almond trees need cool winters and hot summers and do not thrive in below zero conditions, and that is what exactly the climate of California is. Bees are required for pollination as the almond trees are not self pollinating. The almond trees are in full bloom in February and it is during this time that bee keepers set up their bee hives in the almond orchards so that the harvest may be lucrative for the almond growers. Because of the inability of almond trees to self pollinate, the almond growers plant several varieties of almond trees.

Harvesting of almonds

Almonds are harvested after August and in the beginning of October. Almonds have to be harvested when the shell splits and the nut dries up completely, because of this opening in the shell.

When the almonds are ready to be harvested, the orchard owners have the areas swept clean of all debris so that the area is completely clean. Once this has been done they call in the mechanical tree shakers which shake the trees gently and all the dry almonds fall to the ground. They are left there to dry up completely. The well dried almonds are swept into rows and are deposited in a huller after being collected mechanically.

Almonds are high in nutrition

Almonds are a highly nutritious food as they contain vitamin E, Magnesium, Riboflavin, Copper and Phosphorous. At the same time it is very low in fat content and is free of cholesterol and sodium which are bad for the heart. More than half of the almonds cultivated in California are exported.

Beekeeping In Different Areas Of The World

Several locations all over the world produce honey for food and medicine. There are many beekeepers in certain parts of Europe, Africa, Asia and the United States. Though bee keeping started in Europe and was brought to the United States later, it is more modernized in the United States. There are different cultures for keeping bees in different countries. Honey is used as food in most of the cultures and in some for religious concoctions that are sweetened with honey. The Americans make honey for commercial purposes to be sold in supermarkets and also to send it abroad to other countries for sale. The honey is put into containers with their brand names put on it and sent out to areas where honey is scarce or not produced at all.

There is no mass production of honey in many countries

Unlike in America, most of the other countries do not produce enough honey to be able to ship some to other countries too. America manages to harvest enough honey in one season so that it lasts them until next spring to get ready for the next shipment. The bees are usually inactive during the winter and produce honey only in the warmer summer months when there are more flowers around the place for pollination. This is when the mating starts too and then the bees are busy producing honey and taking care of the larvae. United States is known to produce the most honey around the world and also to provide it to the supermarkets all over the world.

Maintaining of hives

The hives do not need too long a time on a daily basis for maintenance. If the beekeepers watch over the hives one hour on a daily basis this should be sufficient. This is needed mainly during the peak season which is between May to September. If the season is good the beekeeper can get around 60 to 100 pounds of honey. What also matters is the price that the beekeeper gets per pound of his honey and with this he can gauge what his earning will be in the future for every harvest.

How to increase the produce of honey

The biggest problems the beekeepers face are the bumble bees which hide in the ground and swarm up from there to visit the flowers that have already been visited by the bees. To avoid this beekeepers changes the place of the hives around, and this is also beneficial to the honey production as the bees have new flowers to pollinate and more honey is cultivated this way. This is called the migratory system amongst the beekeepers when they change the place of their hives. The different batches of honey produced with this vary according to the pollination. The taste of the honey depends on the flowers that the bees have been pollinating.

Beekeeping And The Apple Orchards

The different ways apples are consumed

Apple orchards are all over the country and this is where the super markets procure their apples from. Apples are a very popular fruit and products made from apples are also widely consumed. There is apple sauce, apple cider, apple juice and then there are baked items with apples like apple pies and apple cakes. Apples are a favorite fruit of many and are good for all kinds of health conditions with no restrictions because of medical reasons on this fruit. It would definitely have been a very different incomplete world without apples!

Fall time is apple season

The apple tree looks picturesque and beautiful in every season, with apple blossoms filling the apple trees in spring, with their sweet fragrance; in summer the tree is full of lush green leaves; the tree is fully laden and in the winter the branches which are spreading far and wide are covered with layers of snow and in fall the juicy apples glisten in the sun ready to be picked. People who have been admiring this beautiful tree are also taken up with an other aspect, which is that they do not see anyone working in the orchards right through the year expect when the trees are laden with fruit and at that time it is to pick the fruit. This shows them that there is really not much work right through the year except at fruit picking time! With this in mind apple orchards are bought up as soon as they are put up for sale.

An apple orchard requires a lot of care in reality

Actually an apple orchard, unlike the impression that people have, is a lot of hard work. It is really not easy money as people imagine it to be. These trees have to pruned regularly; they have to sprayed to protect them from pests; a lot of maintenance work is required as the older unproductive trees have to be removed and new trees have to be planted in their place. So there is a lot of work that goes into maintaining an apple orchard and just looking at the end result isn't all there is to it.

A larger orchard means larger gains

Experts in agricultural business feel that an apple orchard should be a minimum ten acres to yield good returns and at least break even. So any new buyer has to think about the size of the orchard when he plans on getting himself an apple orchard. A larger orchard would mean better returns but it also means more expenses as more equipment would be needed, more helping hands would have to be hired, there would be more new trees being planted in place of older ones that have stopped yielding fruits and then more pesticides and nutrients would also be required for a larger area. So for better profits there would have to be larger investments by way of time, money and effort.

Newcomers should not invest in an apple orchard during spring

Spring time when the apples blossoms are in bloom is definitely the wrong time for new comers to buy an orchard. This is the time when the flowers have to be pollinated for the tree to bear fruit. Though the flowers do get pollinated with the wind and birds and other insects, there is nothing like the hard working bee to do a much more effective job of pollinating the flowers. It would not do to depend on the bees in that locality to carry out the pollination; instead it would be much better to hire out the services of bees from local bee hive owners. These owners will set up the hives in the orchards and the pollination will be much faster and much more effective.

Acquiring The Bees

Bee Keeping A Hobby And A Business

Summary: Bee keeping can be a business venture or a hobby. For those who would like to do something in the agricultural field and do not have too much place or money this would be an ideal agricultural business to star. Of course after ensuring that they are not allergic to bee stings!

What it takes to start bee keeping

Bee keeping is not a very expensive venture and for anyone with a little space it is possible to start this business with just about $300 for one hive. This is enough to get started with and you can expand later if space permits. However, the very first thing that you should do is to find out whether you or any family member of yours is allergic to bee stings, if not you can go ahead and get your first bee hive. Another point that you should clarify from the local cooperative office, is whether you are permitted to keep bee hives in your area. If you can do so, then you will have to get registered as a beekeeper with the beekeeping organization.

Select a suitable place to cultivate your bees

Once you have your beehive, you can keep it away from the home somewhere in the corner of your back yard in a spot which you feel will be suitable to have your beehive.

You will have to have the necessary equipment for maintaining your beehives successfully. You can find out about the equipment that you will need from the Federation of American Beekeepers or Cooperative Extension office of your locality. Most of the equipment is available online and through EBay. You could search on the internet and get all the information and equipment that you need and order it to be mailed to you.

Play it safe with bee hives

You will have to ensure that you have the proper beekeepers tackle and do not get stung by bees. Make sure that all those who will be handling the bees have this gear available to them whenever required.

Place an order for your bees at an apiary

Order your bees only after you have all the equipment and the bee hive in place. The apiary where you order the bees from should be one that is well established. Winter is usually the bet time to order your bees, which means around January or February and you will get your shipment of bees by March or April. The U S postal service carries the bees for apiaries and will inform you once the bees arrive so that you can come and pick them up. Mail carriers do not like to carry a box full of bees that are agitated with the journey all over the countryside. You should pick up the bees as soon as possible as they are not going to remain healthy for too long in the confines of a vehicle.

Bees are packed in a "House"

Bees are normally shipped in wooden cases specially made for this purpose. The package will have a sign saying "house" on it. These wooden frames are covered with a screen which allows air circulation and also protects those who have to handle this package like the people from the postal service, from being stung.

There will usually be a few dead bees lying on the bottom of the container, but do not get perturbed. This usually happens as you cannot expect all the bees to take the journey well. You will find the remaining bees clutching on to the sides of their container.

The queen bee

The queen bee is kept separately along with a few nurse bees, and her container will be covered with a piece of sugar candy. The rest of the bees for your bee hive will be put in a container together and these will form the remaining hierarchy. In this container there will be a bowl of sugar syrup which is for the bees to feed on while traveling. The bees will need a drink once you get them home, for this you should spray the container with a fine spray of water. Now your bees are home and you can get started on your bee hive.

Beekeeper Training – An Industry In The Making!

Acquiring training in bee keeping is something that cannot be achieved overnight. One needs to first learn all about the habits of the honey bee in order to cultivate them and safely too. If you do not know what you are dealing with you could be in for some very painful hours nursing bee stings.

Beekeeping training includes not only learning how to attract or acquire your swarm of bees but how to keep them safely so as not to endanger your neighbors and passers by. You must also learn what clothes you need to wear while dealing with your bees and the various actions that you need to take during an emergency.

Bees are very temperamental creatures and if they are disturbed or feel their hive is danger any moving object will be a target for them and they will swarm down in the thousands to attack the moving object, invariably humans. People have been known to die from bee attacks.

Your bee keeping training will also teach you how to administer first aid to bee attack victims and all this is apart form teaching you how to attract bees, encourage them to produce honey, safely remove the honey, and process the honey. Training will also include hoe to market your honey.

Bee keeping training is usually imparted with the help of large honey marketing firms. They want small bee keepers to harvest honey and process the honey which they buy and market all over the world.

Bee keeping is not all about constructing a wooden box and placing it for bees to begin producing honey for you. That is easier said than done. Beekeepers need to know exactly where to place the hive for bees to set up their home in it. There are techniques that attract bees to the hive. One of the best ways is using a metronome to produce a humming sound that attracts the bees. Once the bees ate in he hive the metronome is switched off and he bees decide to make the place their home.

Good bee keeping training will make excellent farmers. Bee keepers need to know everything about honey bees, their nature, their lifestyle and how they search out sources of pollen for their honey. Many bee keepers are taught how to cultivate certain varieties of flowers where the bees can collect nectar and store it in their hives.

Bees are most inactive during the warm summer months and live on the honey that they make in the winter time. However, bees will continue to make honey until their hive is not substantially stored with honey. So, if a farmer is trained properly he will know how to remove the honeycombs and remove the honey. This will fool the bees into thinking they have not made enough honey and will then go out make some more. However this can only be done in the cold winter months. So, a farmer must know how to time the harvesting of honey from the hive.

It's amazing that bees have mastered the art of survival during the winter months. Beekeepers also have to keep in mind that certain times of the year there may not be any honey production since bees are most active during the warm months so that's why many of them are actually farmers since they have to have a way to make a living when it gets cold. This is an expensive hobby and it may look cheap because you can make a box put some slides in them and allow the bees to come there, but the thing is that you have to know where to put the boxes for the bees to build their hives in.

You have to train yourself to be knowledgeable in the area of entomology because you have to know what insects will be compatible around bees because some insects will feed on bees, yellow jackets, hornets, and wasps which are primarily mites and are one of the most annoying insects because they're so relatively tiny that you need a microscope to see them up close. Science plays a huge part in a beekeeper's training and gaining experience since most people aren't savvy to science and the elements of it which is important and necessary because you have to have some idea of how to manage bees and what to do to keep their habitat healthy and to keep pests from overtaking the hives and killing the bees. There are a lot of steps involving the proper education and training of a beekeeper and what you're looking for is someone who is serious and dedicated to a way of life that's been a tradition in some families for generations.

Many people learn through the ranks of great grandparents, grandparents, and parents and it's just a family tradition and way of life that's taught to children. It wasn't even about making money it was actually just one other chore on the farm, but in the years it slowly progressed into a farm staple that was being sold like it was produce, meat and dairy, but it's still a profitable market anyway you look at it and it's one of the sweetest things in the world.

Queen Bee – The Key To Survival Of The Hive

Scientists have studied the nature and habits of the honey bee for years and have some explanation to offer regarding the creation, nurturing and reproduction habits of the queen bee.

Every bee hive needs a queen bee as the very survival of the hive depends on her. Each hive will have one living queen bee that has just one duty and one duty only, that is to go about laying eggs in each larvae cell in the hive. This process goes on for the life of the queen bee.

When the workers determine a suitable place to build their hive they begin to build a few cells where the queen bee begins to lay her eggs. Each group of bees has their work cut out for them. One group will assist in the nurturing and feeding of the larvae, one group will fly out to scout for pollen while another will collect the nectar for the hive, and still another group will change the nectar to honey.

While the queen bee lays her eggs the other bees nurture the larvae. It is during the first 3 days of incubation when the worker bees determine which of the larvae must be nurtured into queen bees. There will only be one queen bee leaving her cell. Unlike the other cells in a bee hive the cells that house the queen bee larvae are vertical instead of horizontal. The worker bees feed these larvae with a substance from their heads to turn them into queen bees.

Some of the larvae in the other cells are nurtured into drones – to mate with the queen bee when she swarms. When the queen bee is ready to hatch she begins to emit a very unique sound that warns the other workers that the new queen is emerging. The workers and the other bees developing in their cells will leave with the new queen in swarms of thousands. The queen bee will fly high up in the sky while the drones will try to catch up with her and mate with her. The first to reach her will mate and then die. This is the first and last time the queen will swarm.

When the mating process is over the workers will kill the remaining 'virgin queens' in their cells. They will not harm the remaining queens until the emerged queen has successfully mated.

The queen after mating will retreat to a tree where she will be engulfed by thousands of workers who strive to protect her. When the others have found a suitable place to build the hive she will proceed to begin to lay her eggs. In the meantime the first queen proceeds to lay her own eggs.

History of Beekeeping

No one can pin a date on when humans learnt to cultivate bees for their honey. Neither can anyone say for sure when man learnt that honey can be consumed. Perhaps man observed the bears and other animals in the trees pursuing honey bee hives and taking the honey from it and eating it. And as always man – following his 'monkey see, monkey do' instinct tried it out for taste and discovered that honey had a unique sweet taste of its own.

All this probably happened around 700 years before Christ. We can fix this date by the drawings on the caves where prehistoric man lived. It is not difficult for scientists to fix dates on caves based on radio carbon dating of skins and bones along with other organic remnants found in the cave. However, scientists seem to think that the drawings depicted on the walls of the caves tell tales of humans standing around a bee hive and are absolutely unaffected by the bees. Could this mean that they had discovered some pre historic 'Bee Go' to ward off the bees. Bee Go is a chemical that bee keepers smear on a tray and leave it in the hive to encourage the bees to move to the bottom of the hive.

Man might have discovered that smearing some similar substance on their bodies would prevent the bees from stinging them while they collected the honey from the hive. What is a fact is, however, that man did learn that smoke was a very useful tool in scaring the bees away from the hive while they dismantled it from the tree. This was no guarantee of not getting stung though.

As civilization progressed man learnt to cultivate the honey bee. First indications of bee keeping are the clay and mud pottery scientists have found. Artificial bee hives had been constructed out of clay pottery to encourage bees to build their hives in them. When the honey was ready for harvesting the bees were driven out and the pot was broken to remove the honey.

Swarms of bees were trapped in these containers and kept there captive. Once they decided to make the hive their home the bees were allowed to come and go as they pleased making honey in the process.

Time advanced and so did the human intelligence. Man began to study the habits of bees and made uniquely designed hives for the honey bee. These wooden hives as we know them today are engineered to keep the reproduction part of the hive away from the food (larder) section. This way the bee keeper can only remove the honey combs without disturbing the off spring at the same time not having to remove all the bee's food.

Swarming – A Natural Process Of Mating!

Come spring and you will notice honey bees swarming. Swarming is the process when bees, in the thousands, will fly with the queen bee in the attempt to mate with her and also find a new place they can build their home.

The bee hive is a very organized place. Each group of bees has their work cut out for them. There are the worker bees that need to build the bee hive, there are the bees that scout for new locations for hives and sources of nectar and there are the bees that server the queen bee that is only concerned with laying eggs in the cells of the hive.

The hive is divided into cells that are demarked for honey and larvae. The queen bee goes about the hive depositing eggs in the cells and the worker bees cover the cell with wax to protect the larvae. It is surprising how the sex of the larvae is determined by the way the worker bee nurtures the young.

Some of the larvae will grow to be worker bees and some will grow to be females, which are searched out and killed by the queen bee. There can only be one queen bee in the hive. The worker bees will strive to save the female bees as they have to reproduce, these worker bees will do all they can to prevent the queen from killing all the female bees.

One of two things will happen in the spring every year. Either the workers will drive out the queen bee along with thousands of her workers, or the queen and her workers will drive out thousands of workers who will leave with one queen bee. This flying away in thousands is known as swarming and so a group of bees is called a swarm. They are actually trying to mate with the virgin queen and then look for a place to build their hive.

When bees swarm, the queen bee flies very high and a group of bees called drones try to catch up with her, while the workers fly in pursuit. The first drone to catch up with the queen bee mates with her and then dies. Once the mating is done the queen will retreat to a safe place and will be surrounded by thousands of her workers who protect her.

When you see something that looks like a hive on a tree it is actually hundreds of thousands of worker bees surrounding the queen bee protecting her from possible attack, while the scouts have gone out in search of a suitable place to build their hive. At this time they have no honey to sustain them and are very hungry.

Some bee keepers are very good at attracting swarms of bees to their man made hives where they cultivate their honey making skills to the maximum.

Processing Of Raw Honey

When the bee keeper removes the honey from the honey combs he has to process the raw honey immediately to prevent it from crystallizing. Once the row honey comes into contact with the oxygen in the air it reacts and begins to crystallize immediately. This does not happen in the honey comb as the wax caps over each cell in the comb keeps the honey away from the air.

Apart from processing the raw honey to prevent crystallization, the bee keeper has to kill some very potent bacterium in the honey to prevent poisoning. This is the bacteria that cause the symptoms of botulism in humans. These symptoms are better known as 'food poisoning' and are removed by heating the raw honey for some time at temperatures between 150 to 170 degrees centigrade.

Honey is naturally sweater than processed sugar and is stored in its natural color. Sugar made from sugarcane is bleached to remove the natural brown color of sugar. This is why table sugar is white and crystallized.

Honey is a whitish substance that is very gooey in nature in its raw form. It is only the pasteurizing process that causes it to get that yellowish color. When the honey is processed on a hot fire it begins to caramelize, very much the same way sugar does. Honey has a very long shelf life and cam be kept for years after processing. People buy honey for it's medicinal value as it has many vitamins and an extremely high amount of antioxidants and digestive enzymes. The healthy properties of honey cannot be enumerated in this short article.

Honey is fast taking its rightful place in society by replacing many substances that aggravate diabetes. For instance corn syrup is being replaced by honey as corn syrup is a known cause of diabetes. Corn syrup is a product of man through many automated and mechanical processes where as honey is only processed to remove bacteria from its constituents. Any sweet produced by man is a sure cause of diabetes. Natural sugars are not.

Honey has a lot of medicinal values too. For instance it is used as a topical application to treat conditions like MRSA, this is a type skin infection. Honey is also known to be very good for the treatment of laryngitis. A bit of honey mixed with a bit of lemon will sooth your throat as it does contagious conjunctivitis.

Beekeepers across the world like to stick to the organic way of processing honey because they do not believe in using harmful chemicals to purify their produce. These chemicals if used will destroy the many benefits of honey in its natural form. This is something the large manufacturers of honey cannot guarantee, many of them do use harmful chemicals to process the honey they sell. This is the reason people stick to local farmers for their supply of honey.

How To Market Your Honey

If you are an upstart and just setting out into the honey business you should weight the pros and cons of marketing the honey you produce on your honey farm. There are many competitors in the market today and they have access to the best marketing services. Their honey is also packed in various size containers that give a customer a choice, something you may not be able to offer.

Large businesses have invested a lot in marketing their honey. They have set up a very effective advertising system, a distinct advantage over a small time honey producer. However since these people have set up such an elaborate marketing system they have to produce much larger quantities of honey AND sell it in order to make a profit. Here is where the advantage lies for you – a small honey business man.

You do not have to have large over heads to produce your honey, you also don't have to cater to a world wide market. You can concentrate on the local stores and neighborhood buyers to sell part or even your entire produce. It may sound a bit astounding but your bee hive can produce over a hundred pounds of honey. SO if you have three of four bee hives you can rake in a sizeable income by targeting a small group of customers.

There are a few techniques of marketing your honey.

1. First of all there is the option of purchasing small glass or plastic bottles and customizing them with labels that you can get printed at any local store for quite a reasonable price. Fill these bottles with your honey and fix up with local stores to sell them for you , at a profit.

2. You could set up a curbside honey stall and sell the honey to passers by. People love fresh honey and will definitely stop to buy some from a curb side stall as this is a sure shot way of getting some fresh honey. Honey from the store may be months old by the time they purchase it. People are, however, skeptical of buying large amounts of honey from curbside stalls as they are afraid of the honey crystallizing, it is up to you to convince them.

3. You could also advertise in a local publication and circulate the availability of your product by word of mouth at local school meetings and clubs. You will be surprised at the response you will get.

4. Then there is the option of contacting a large honey marketing firm to purchase the honey from you as most of the marketing firms buy their honey from bee keepers and market it under their own brand name. This is probably the best option to sell your honey.

Honey – Natures Sweetness Delivered By The Bees!

Honey bees work tirelessly to assist in pollinating flowers in the fields and collecting the nectar in their hives in order to process it into the sweet tasting honey as we know it. The nectar bees collect from the center of the flowers and process it to make a thick sweet liquid that is what we call honey. This honey is nothing but food for the bees.

In the process of collecting nectar from the flowers the bees help nature pollinate other flowers as the pollen from flowers is distributed or delivered to other flowers when the bees fly from one flower to the other. This is how the reproduction of flowers is assisted by the honey bee.

Bees move in swarms and these thousands of little creatures work tirelessly to collect nectar and deposit it in their hives where another group of bees process the nectar and turn it into honey. This honey is stored in special chambers in the hive called 'honey combs'. The bees feed on this honey.

Humans have developed a taste for honey, which is very nutritional and has some medicinal value too. So they have learnt to cultivate honey bees in man made bee hives. The bee hives are constructed out of wood and make it easier for the bee keeper to collect the honey as and when it is ready without disturbing the bees and their off spring.

The man made bee hive looks like a wooden box no larger than 1 and a ½ feet by 1 and a ½ feet and stands about 2 feet tall. This four sided structure has a small opening from where the bees enter and leave the hive. Once the bee hive is set up and placed on a flat surface to prevent it from tipping over with the wind, the bee keeper may introduce bees to the hive by buying a swarm from a bee farm or he can just wait for the bees to find the hive themselves. Both ways work equally well.

Within a matter of 40 days the bee keeper will be able to slide out the honey combs without disturbing the bees. Care must be taken not to remove all the honey or the bees will starve. This way, the bee keeper collects his keep for providing the bees with a safe place to breed. Cultivating honey bee farms is a very good way of earning a living and a fun way too. However, some sort of training is required, mainly for the safety of the bee keeper as bee stings can be fatal at times – so a bit of caution and safety techniques is required.

Harvesting Honey

Now every one knows that you are not doing nature a service by providing the bees a place to live by setting up a bee hive, they are quite capable of doing that themselves. Your main intention is to harvest honey, and a lot of it. The best way is to set up a bee hive and regularly inspect it for the produce.

By inspecting the supers in the hive you will know that it is time to collect your share of the honey hive when you notice the supers have honey comps and are closed with caps of wax. All you need to do is to take out the honey combs and get to the honey – easier said than done!

Now harvesting the honey fro the bee hive will not be such a problem for the experienced bee keeper. You will need to wear special be keeping gear that will prevent you from getting stung by the ferocious little creatures that can get pretty aggressive if someone tries to steal their food.

Bee keeping gear consists of light colored clothes, because bees are attracted to bright colors. You must also stay calm if they swarm over your face mask. You will also need some additional tools such as a scraping tool and a smoker.

When you are sure the supers are full, you can proceed to encourage the bees to leave the super. Some chemicals available in the market will make this task easier. One very popular chemical that is used to scare the bees away is called 'bee go'. This is applied to what is called a fumer board when the bees get a scent of the 'Bee Go' the bees move to the base of the hive. This leaves the hive free from any bees ready for you to take out the honey combs. Fisher Bee Quick is another good chemical that assists in removing he bees from the hive without harming them. They just find the scent very offensive and move to the bottom of the hive.

Once you have safely removed the honey filled honeycombs from the hive you need to extract the honey from it. You must first remove the wax caps from the cells. These wax caps seal the honey within the combs. You can use a metal knife to remove the wax caps, this is better achieved if the knife has been slightly heated on a fire as it melts the wax a little. It is, in fact, better to warm the knife by dipping it into a basin of hot water.

The honey will begin to drip from the comb once the caps are removed. It is best that you place the comb on a cheese cloth that has been placed over a pot to collect the honey. The honey will strain through the cloth leaving the caps behind.

Curbside Honey Sales

You will find the curb side honey stand just as common in the rural parts of the United States as you will find the Lemonade stand in the city. It is common for the American farmer to set up a roadside stand to sell off his surplus produce for the season. This is a very effective way to sell off and make some money off the extra vegetables and fruit produced in the farms in the countryside. These same stalls selling farm produce can be used to sell all the honey you have collected from your bee hives. Setting up and selling honey in this fashion is called curb side honey sales.

It is important to set up a large sign on your curb side honey stall indicating that you are selling fresh honey from your bee hive. Make the sign as simple and precise as you possibly can. Use ink or paint to write on the sign and make sure it contrasts with the background of the sign. Remember that people drive by very fast and the sign must be large enough for them to be able to read from at least 50 feet away. Remember you sign should be clear and short. Desist from drawing on your sign. Drawing bees on your sign might have people thinking you are warning them of the presence of hostile bees in the area. This may not assist you in your sales endeavor.

Examine your honey for floating dust and other remnants of the hive before setting it up for sale. Make doubly that you have strained your honey before trying to sell it on your curb side honey stand. Any signs of dirt or wax from the hive can put a customer off and he or she will never return. Hygiene is the key to getting customers to return to your sale stand. As with the inside of the container and the contents you should take pains to keep the outside of the containers clean as well. Remove any signs of dripping honey from the rim or the outside of the container and then dry the container so that dust will not stick to it on the curb side.

You should, if possible, offer you customers a variety of honey in a variety of jars. This means you should have many sizes of honey jars they can choose to purchase. Many people prefer to buy large containers of honey because they are not sure the honey will not crystallize. A good idea is to decorate your stall with flowers, or perhaps you could even sell the flowers along with your honey.

If you have a shaded section of your front yard this is the best place to set up your honey stand. Try to sell some food stuff along side your honey, which is your main product. Sweet corn, banana peppers and more could really be an added attraction. The shaded part of the lawn will make your products look like they are farm fresh and attract a lot of passers by.

You should also develop a salesman's spirit and mingle with your customers. This is one way of ensuring that your customers will return time and again.

Beekeeping Equipment

Bee keeping is a hobby, believe it or not. These little stinging creatures can be a lot of fun if handled with care and there are a lot of sweet returns too. However, if you intend to take up this hobby you are best advised to get some basic knowledge about bee keeping - and the necessary mandatory equipment too if you do not want the post office calling you to collect your bees from their vicinity.

Now, the most basic piece of equipment that is required for bee keeping is the bee hive itself. Let us take some time to understand the structure and the necessity of the bee hive. It is not like you are required to go out and get yourself a natural bee hive from out of a tree. This is not at all recommended as these little creatures are very possessive of their home and the occupants so you may just as well land yourself in bed with multiple stings that may also prove fatal at times. The hive we are talking about is constructed out of wood and looks like a small wooden cabinet to be kept outdoors.

The beehive you need should have 5 supers. These are the most important part of a bee hive as this is the part of the hive that the bees use to store their product - honey. These 5 supers are placed between the base of the hive and the top cover. Each of the 5 supers contain 9 to 10 frames where the bees keep their off spring and honey. You decide if you want to have deep supers or shallow supers. Deep supers mean that you will have to get yourself a one - size foundation unlike the shallow super where you will need multiple size foundations. The bad part of a deep super is that you will have to lug out a hundred pound weight when it is full.

When you are ready to set up your hive ensure that you place it on a flat surface where it will be pretty difficult for the hive to tip over when a strong wind blows. Also place it in a place where people and animals will not be able to reach it and disturb the bees.

Now for something called a spacer. these are bits of equipment or rather wooden planks that are used to space out the frames in the super. You will need a few of these but do not worry a beehive kit will usually have all that you will require to set up your bee hive.

Now when the bees have created enough honey and you need to get it out of the hive you will need to have a 'smoker'. This is not a cigarette puffing human but a piece of bee keeping equipment that is used to 'smoke' the bees away from the hive so that you can safely collect the produce inside the supers. Smokers are simple in design and are constructed out of a funnel and bellows. You will need to use some smoking material such as 'Brlap' to create a lot of smoke, dried corn cobs are another good alternative to create a lot of smoke.

Your bee hive will come with a metal hive tool used to pry open the beehive and scrape the honey from the frames. A furmer board is used to encourage the bees to leave a super and let you take their honeycombs. Now for the bees themselves, You could order them or set up your bee hive and wait patiently until they find it and build their colony themselves.

Go to the net and find the best way to order your bee hive and also the inmates. This may be the start of a very interesting hobby.

Beekeeping Companies Are Owned By Families Down The Line

Beekeeping supplies

Companies that have beekeeping stuff deal with all the equipment that is required for this business, like attire for bee keeping which is essential from head to torso, full body suits and just head gear. Along with this equipment they also sell journals and books on beekeeping to help people to understand this field better. Some of the better known beekeeping companies have been in the business for more than a hundred years.

Books for providing short courses in beekeeping

Beekeeping families have a lot of subsidiaries to the beekeeping business too. They use the bees wax to make candles and collect honey which is sold locally in the stores. The books have information on candle making and collecting honey too apart from instructions on the main business which is bee keeping. These family businesses are mostly long standing ones and continue from one generation to the other.

Businesses have advanced technologically

Stepping into the shoes of their ancestors has not been easy for many of the families. Only those families where the parents thought of the new technologies and got their business mechanized with the latest equipment, and advertised with on the internet with their own websites so that they could bring in more customers and higher profits, was the business profitable and worth running. For others the cost of maintaining an old business was too high and they could not handle the expenses anymore. As most descendants of beekeeping families knew no other business they preferred to stick to it, and if the financial situation was not good, then they became commercially owned businesses

A business that once started as a hobby, with just about enough honey to put on toast and biscuits in place of marmalade, became a big time business with the popularity of the use of honey. Earlier honey was not a very popular edible thing as people used sugar and molasses for sweetening their food, but once they found that honey was much cheaper, beekeeping became popular and was taken up by many as an agricultural business. However, now the only problem is that people worry about the safety of the honey that is collected because of all the pesticides that are used. Now with organic foods being more popular and a movement by organic food growers, there is a lot of honey that is free of pesticides. Now many families owned companies supply bees wax for candle making and cosmetics to other manufacturers of these commodities.

History Of Beekeeping

Bee keeping dates back to the 13th century BC when it was practiced by ancient Egyptians. Once it was introduced by John Harbison to the United States, it became more modernized with updated techniques and became the main profession and financial sustenance for beekeepers. There are other by products that are derived from honey which is propolis and royal jelly which are used for medicinal purposes. Not much has changed since ancient times in the use of products derived from beehives.

Different varieties of bees brought into the US

Bees of different species were brought from various countries like New Zealand and Europe. This was more of a hobby for those who lived on farms than a main means of earning their keep. Farming was the main occupation and beekeeping was just a side hobby that was followed by the relatives of the farmers, who had access to space on the farm and found it easy to keep bees. This hobby then was passed from generation to generation.

From honey and bee wax the bee hive changed with science

In early days, bees were maintained just for their honey and for the bee's wax that they produced for their hives, and which was used for making candles and various other products. Later L. L. Langstroth an American scientist brought more scientific methods into bee keeping and brought into practice the beehive frame that was removable. Later it was found that bees could be influenced into building their own frames that were straight by giving them some wax as a foundation. The bees would then make use of this foundation and build their own honey comb with holes that were octagonal in shape to keep their larvae in until they had developed enough and hatched. The methods for beekeeping kept getting more and more developed and a helpful and practical invention was the smoker, which assisted the beekeepers as a safety device.

The art of beekeeping

For any successful beekeeper, this art should be second nature to them and they should know all that there is about bee keeping. It is always easier for those who are born into beekeeping families to work on such projects as this has been their life from the time they were born. With beekeeping being the family business the new generation will have no problem in picking up the strings as easily as it would be to learn how to walk and talk!

Apiculturists are agriculturists who are bee keepers

Bee keepers are agriculturists of sorts as their profession is closely related to the farmer's profession, who breed cows and grow food side by side on the same farm. The same way, many farmers have bee keeping as an additional source of income and a hobby for the others in the family to carry on some trade and earn some extra money. Of course bee keeping is also a full time profession for some. The Department of agriculture refers to bee keepers as apiculturists.

Beekeepers who know about entomology and biology are more successful in their bee keeping business and can advice those who are in need of some more know how on bee keeping. They can help many a bee keeper with their knowledge and if they pass it on to other apiculturists they will be helping them with their business too.

Honey Extractors

How to make your own honey extractor

To extract honey from the beehive one must have an extractor of honey. Honey extractors are available in the market and cost around $300, which is the approximate price of getting a new bee hive. Sometimes instead of investing so much money in a new honey extractor, several beekeepers who are located near each other collectively get a honey extractor and share it. In case there are no beekeepers near your place, then the next best option is to make your own honey extractor.

How to make your own honey extractor

To put together your own honey extractor you would first have to get all the materials that would be required for this. What you would need are six coach screws, is a single pillow block bearing, a bearing that is self centering, four sections of 400 mm of 8mm rod that is threaded, for the metal bolt you will require ten bolts, a metal drum that is fairly large, a single meter of fencing wire which is 2 to 3 mm thick, two bits of wood, two wheel rims of a bicycle and a one meter rod of metal that is thickly threaded. While selecting the metal drum that is a large one for storing the honey, make sure that no toxic or dangerous material was ever stored in it before. The tools that you will require for making your honey extractor would be a hack saw, a socket set, a machine for welding, and an electric drill.

The step by step instructions for making the extractor

See the side of the drum that is without the two holes and open it, this will be the upper end of your drum for the honey extractor. With the coach screws fasten one piece of wood diagonally across the underneath of the drum. Now put the pillow block into place securely with the coach screws.

The threaded rod should be inserted through the middle of the initial bicycle rim, firmly lock the frame to the rod about ten centimeters away from the rods end. On the opposite side of the rod thread a but for the other wheel, on this nut you should make the second wheel will rest. Drill four holes and when both the wheel rims are in place around each wheel. After this job is done use the 8mm rods to secure the wheel rims together. Using two of the nuts on the rod, ensure that two cm of the rod stick out

Next you should cut a slit 10mm deep and 3mm wide in the last part of the rod. After this thread the lock the nuts together at the finishing end of the rod. Once the nuts are in place, with the welding machine permanently secure them into place. Secure the wire to the spokes of the wheel rim which is at the bottom, about 5-8cm from the edge of the rod. Your honey extractor's basket is ready. Take this basket and put it inside the drum on the pillow bearing. Now bolt another bit of wood to the self centering bearing and the sides of the drum. Drill a screwdriver piece into the chuck and place the chuck inside the slit at the slot on top of the threaded rod. For details with photographs for a better understanding of making an extractor for honey go to the websites for more information.

Packaging Your Honey

Packaging honey for commercial use is an important aspect as this has to be according to the USDA standards. To be able to market the product well and for it to be an attractive shelf item, the packaging is the most important aspect.

How large beekeepers can make a profit

To be able to make a good profit large beekeepers cannot just sell their products within the local community but will have to spread their wings and get business from the super markets and grocery stores too. Only then can they cope with the financial expenditure and remain solvent financially. For this to happen the beekeepers have to pay heed to the packaging of the honey and other by products which should meet the standards set by USDA.

How do they decide on the packaging?

The container is what makes the package attractive and this is what the beekeeper has to bear in mind when thinking of the packing. These packings can be in various types of containers like glass bottles, plastic containers and cans. The sizes of the containers can vary from a few hundred grams to several kilos. The smaller containers come in attractive shapes and colors and can be reused for storing anything else by the buyer later. Another aspect is to keep the bottles and containers firmly sealed so that they can be shipped to any destination without fear of leakage.

Labels are equally important as a visual effect

After the container has been decided on you will have to think about the label which is what makes the container colorful and attractive and gives the buyer details about the product. Before designing the label or going to an artist who will help you with the design you should check with the government of your state about the laws that govern certain requirements. This will decide about the information that you are supposed to put on every label. The name of the product, which in this case is "Honey", should be mentioned boldly on the label. If you are using a distribution or packaging company their name and address as well as the name and address of your farm should also be on the label. Apart from all this the date of packaging the honey and the net weight should also be mentioned clearly. The size of the font on the label will be according to the size of the container and the label.

For those beekeepers who harvest honey of different flavors, the name of the flavor should also be mentioned on the label, as different people would prefer different flavors. If the honey is not filtered then you would have to mention that it is raw, natural or unfiltered on the label too.

USDA grades

The beekeepers who have the USDA certification will also have their honey graded and the grades will have to be mentioned on the labels too. These grades are based on the defects, quality of flavor, clarity and the amount of moisture in the honey.

Marketing The Honey Locally

Beekeepers who have a small scale beehive business sell their products locally to friends, neighbors and known people. The marketing for their products is done by word of mouth.

Advertising beekeepers produce

Initially some beekeepers start their business on a small scale and at this time they sell their products locally. Some of them even set up a wayside stall and put up some of the other products from their farms like apples, and other fruits and vegetables for sale. This way they create a market for their products and their stuff gets advertised by word of mouth. If the sales increase and people find that the quality of their products is good, then they could even start putting their stuff in the local stores and markets too. The beekeepers get to know most of their customers personally as they meet them regularly when they come to buy the honey and other products from their farms and this creates a good rapport between them. So not only do they make a good sale they also end up with good friends.

More ways of increasing sales

The way their wares are packaged is another important aspect of increasing their sales. Making the labels and containers attractive also attracts the customers. The labels should be legible and catch the buyer's eye and the bottles clear so that the honey can be seen in them. The labels should give details of the name and address of the farm where it is produced and the date it was packed on. The bottles should be of a quality that the customer would want to retain and put on his kitchen shelf. The beekeepers should make sure that they keep their stall spotlessly clean and arrange their display attractively too.

Signboards for road side stalls

To make the roadside stall more noticeable the beekeeper should put up a signboard which can be seen clearly by those who are driving by and should mention in bold letters that they have put up farm fresh produce and honey for sale. There should not be a jumble of words and colors on the signboard but it should state the facts in bold and clear letters. A canopy or some sort of sunshade would make the customer more comfortable instead of their standing in the sun. This way they will not be in a hurry to get out of the sun and into the comfort of their cars.

Ensure top quality products on the shelves

Make sure that what you put on the shelves in the stall is always of good quality. If you notice the honey in any bottle crystallizing immediately replace it with a fresh bottle of honey. Do not let anything that is not in good condition be seen by the customers.

Give out pamphlets at the stall

No doubt the roadside stalls do increase the sales of the beekeeper's produce, but to gat a still better response giving out cards and pamphlets would be helpful too. It would be nice for the customers if pamphlets with recipes using honey are given to them and also a list of all the other products that they are selling here. A few small free samples being handed out is another good way to create a good rapport with customers too.

While pricing the honey take into consideration the time and effort spent on the whole job and do not under sell or over price your commodity.

Starting Your Own Beekeeping Business

Beekeeping as a hobby and as a full time business are two different things. This is not only time consuming but also needs a lot of work and effort to be put in to make it pay good dividends.

Beekeeping needs hard work and dedication

For anyone who wants to start a beekeeping business, one thing that they should remember is that it involves a lot of hard work and is not something that is easy to maintain, especially if they want a good profit from this business. Doing it as a hobby may not need as much of an effort as when it is one's livelihood and takes up most of one's time. More money would be invested in beekeeping as a livelihood than if it was just a hobby and if you done this then you definitely need to earn that money back. Not only would you need a good quantity of produce from the beehives you would also need the top quality honey so that your products are sold more in the market. The beekeeper should also get the latest equipment and keep to the latest technology to make sure that his produce of honey is the best in every way.

Most beekeepers have been in the business for a long time and know all the ropes of the trade. They also have a website which advertises their goods and gets them a lot more customers. If you do not have a website then the number of your customers will be very limited as you will not be able to spread the word about your farm products to anyone else apart from those who stay in your locality.

Beekeeping is a competitive business

When compared to the commercial beekeepers the small business beekeepers will have a hard time competing with them. It takes a lot of effort to produce even a small quantity of honey and if the technology and equipment is not the latest in the market then the produce falls even lower and the profit after all the hard word will be negligible at the end of the day. To make the smaller beekeepers remain in business the commercial ones usually take advantage of the situation and buy up or sub contract the produce of the smaller business's and add it on to their own. Other agricultural businesses have co operative societies but bee keeping does not have any such way of helping the smaller beekeepers. However, by subcontracting the smaller beekeepers this is beneficial to both parties.

Sub contracting is not a safe bet

Though subcontracting may sound like a very good idea it is not all that secure as the company that is subcontracting you could suddenly back off if they are not happy with the products or for any other reason. So this could run out to be a risky affair as you cannot tell what he outcome of the business will be and how the season will be for your products. Beekeepers have to worry about their financial gain and keep wondering whether the market demand would be worth while or not in the future. There produce is not as dependable as a farmer who knows what his farm will give him if the climate is suitable and the weather good for his crops.

When you are in the beekeeping business you have to depend on the activities of the bees and how much they produce. This also depends on the weather and the temperatures which have to be suitable for the bees.

The Ecology Of Bees

With more than 20,000 varieties of bees in the world there are various combinations in cross pollination of flowers and the outcome is a different kind of honey that is produced. So the variety of honey is wide and varied.

Cross pollination and the consistency of honey

Beekeepers rely on the several thousand varieties of bees to cross pollinate their flowers and create new species of flowers and also several different consistencies of honey. The beekeepers keep a track of the bees and their cross breeding so that they know where the bees originated from and their origin. This way they also know the consistency of the honey that various bees produce.

Origination of bees

Bees generally originated from Asia, Africa and Europe and were brought to America centuries ago by immigrants from all over the world. The only place where there are no bees is the Antarctica. Bees are related to wasps, but unlike the bees wasps do not pollinate flowers like the other species of the same category which are beetles, butterflies and flies.

The two categories of bees

Beekeepers learn to manage their bees with the two categories which are males and females. Generally there are only a few males and the females that are more in number fight for control of the bee hive with each other.

African bees are not aggressive

People feel a fear when they hear about the African bee and think it is a poisonous killer bee, but this is not so and the African bee is not dangerous at all. These bees are the most popular with the bee keepers and also in most of the beekeeping industry too. Clover honey is produced by the African bee and is the most popular and most utilized honey. These bees never attack anyone, but will do so only if they are defending their hive and the safety of the queen bee is a concern. The queen bee lives inside the hive permanently once she becomes pregnant and will never be seen again. Beekeepers remove parts of the hive but never touch the area where the queen bee lives.

Though bees may be of a passive nature by and large, it is quite annoying to have them buzzing around you while you are out at a picnic. This is because there sense of smell is stronger than their eye sight and they come because of the smell of food. It is their sense of smell that guides them to the flowers and pollinates them. Sometimes the food people consume could smell very much like flowers and the bees go for this. Bees sometimes hover around trash cans because of certain foods that are dumped in them. So farmers should be careful when dumping food in the trash cans because this could lead the bees to the trash cans instead of their natural habitat.

The Life Cycle Of The Honey Bee

If a beekeeper wants to be successful in his endeavor either as a commercial bee keeper or as a hobby he should be able to understand the full life cycle of the bee. This would help him in cultivating and breeding his bees for the best produce of honey.

The unique life cycle of the honey bee

The queen bee lays eggs in the octagonal shaped cells and the eggs get attached to the wall of the cell with a strand of mucus. During spring when the eggs are being laid, the queen bee can lay up to 1900 eggs in a day. The egg remains in the cell until it hatches and there emerges a larva. There are nurse bees that have to care for the larvae and feed the larvae with honey and secretions from the glands which are referred to as bee bread.

There are five stages to the development of a larva, and the larva sheds the outer skin after each of these stages. On the sixth day of this procedure the larva is cocooned by one of the worker bees in its cell where it stays for the next eight to ten days and then emerges as a fully formed bee from its cocoon.

The life span of a honey bee

The life span of the honey bees depends on the purpose that they are in the hive for. Different jobs have different honeybees doing it and their life spans also differ accordingly. The queen bee has a life span of two years but this is only if she has been inseminated with a sufficient amount of sperm while she was on her nuptial flight. The queen bee if she is in a good condition can lay up to 2000 eggs in a day and is also responsible for killing her mother and sisters too. The queen bee has nothing to do as there are enough bees in her entourage to feed her and to clean the waste matter too. An older queen bee will leave the nest while the rest of the bees from the hive are going to swarm. This usually happens during spring time. Those who are proficient in bee keeping, think that the queen bee generates some sort of pheromone that avoids the worker bees from the hives in becoming interested in sex. The virgin queen bee is one who has not had her nuptial flight. The drone bees are the ones that survive only until they have impregnated the queen bee during her nuptial flight. The drone bee dies after mating with the queen bee.

Life of the worker bee

The worker bee lives for one hundred and forty days only during the winter, and during summer it is only forty days. They have a short life span because they literally work themselves to death. There are many different duties that the worker bees have to perform. The nurse bees take care of the larvae while others have to go out and collect pollen to make honey. Other workers have to be capping the honey combs while some have to attend on the queen bee. These worker bees also have to starve the drone bees to death and clean the hives. All worker bees are always sterile and in case they lay an egg are drone bees. Each hive has between twenty thousand to two hundred thousand bees in a one hive. The bees only survive if their queen life, if anything happens to the queen bee then the whole hive dies.

The Science And Technology Of Beekeeping

With modern techniques and scientific developments honey can be cultivated in much larger quantities than before. However, the one draw back is that this is seasonal and according to the work pattern of the bees.

Seasonal activities of the bees

Bees can make honey only during the warmer months when there are a lot more flowers blooming and there is plenty of pollen for them to feed on, but during the cold winter months there is less honey produced by the bees. Bees like humans can sense the changes in weather and environment.

The life of the bees as compared to humans

Bees have lives very much like that of the humans. They organize themselves and their work according to their designations and have a method in all that they do. The only difference between the queen bee and women is that the queen bee is permanently pregnant with her first nuptial flight and stays in her hive producing eggs until the sperm runs out. The queen bee mates with 2 or 3 drones and this impregnates her for the next two or three years. The queen bee lays around 2000 eggs per day and lives for around two years after her period of producing young is over. After this a new queen takes over the hive. Mating is seasonal and the pattern differs from one species of bees to another.

The queen bee, like the dignitaries among the humans, is protected by worker bees and drones who keep buzzing around her to protect her from any kind of harm. These worker bees and drones will give their lives to protect the queen bee. The queen bee's buzz is very different to that of the other bees and is a high pitched buzz. The other bees recognize this sound and keep swarming around her constantly.

Close knit colonies of bees

The entire bee colony is working towards a single goal which is to protect their w queen, bring up the young and make honey. Their final deliverance from all the hard work is their death, which they sometimes bring on themselves by the never ending work that they do. The queen bee also lives only to continue the linage and lay eggs to multiply the numbers in their colony and their hive. Most of the bees in the bee colony are females, but only one will make it as the queen. Beekeepers still find the ways of the bees and their behavior a mystery and never really fathom the life of a bee. Technology and science is trying to learn more about the colonies of bees and maybe learn a lesson or two from them in living together compatibly instead of working at cross purposes in the same family most of the time.

Bee Keeping Safety Equipments

All beekeepers have to protect themselves while they work with their bees. If they get stung by a swarm of bees it could be a fatal attack and so the bee keepers cannot afford to take chances, but have to ensure their own safety always.

Safety equipment for bee keepers

Beekeepers have to wear suits that cover them and protect them completely from head to toe and that cannot be punctured by the bees, and through which the bees cannot sting them. To protect their face they have a mesh screen and also use smokers to calm down bees that have become very agitated. The smoker calms down the bees while the beekeeper gets the honey out of the hive. They also have to check that the bees are in their hives getting on with the honey making procedure and the only way they can do all this is if the bees are calm. Beekeepers also have a crowbar to scrape the honey out which can be quite difficult to remove. The bee hives are usually placed in secluded areas so that the bees do not become a menace to those who are moving around in that area. This is also to ensure that the bees are not disturbed by too much of noise and human traffic which will antagonize them and make them swarm.

Beekeepers get their tools on the internet

Most beekeepers prefer to get their bee keeping equipment on the internet as this way they can access hundred of different companies and get exactly what they want. The need among other things, tools for grafting the comb apart for scraping up the honey, comb cutters, special cages for capturing the queen bee, and many other things like containers for the honey. Honey is the favorite food of grizzly bears and bee keepers have to safe guard their bees from pests which could eat them up. Bees also have to have particular kinds of feeds which keep them healthy. Bee hives also have to have treatment against invasion from flies and moths which carry infection and diseases from compost and animal manure to the hives.

Expenses incurred by the bee keepers

The body suits that the beekeepers wear are fairly cheap, and do not dent the finances of the bee keeper, but what is an expense is keeping the hive safe from pests. This is not an easy task as the hives are situated in places like dense forests where there are many pests and birds to disturb the hives. So the beekeepers have to keep abreast of the new techniques to maintain their beehives in a healthy manner and safe guard them against pests.

How To Transfer Your Bees To Their New Home

Beekeepers have to place their order for their beehives and bees in the winter months. Once the bees arrive the beekeeper should know how to tackle them and transfer them to their new bee hive.

The new beekeeper

The new beekeeper will be more than enthusiastic to have his new bees arrive by post.

Before the bees arrive he would have chosen a suitable place to put his new bees in along with the new hive. The place for the bee hive would have to be in a secluded area where the bees will not be disturbed by both humans and animals. The hive also has to be in a place where strong winds will not knock it down.

Before the bees arrive it would be a good thing to try out the safety equipment and specially the body suit so that you are well protected and do not get stung. The post office will keep you informed on the arrival of the bees, and once they come will ask you to come and take your dangerous parcel away as soon as possible.

The newly arrived bees

In the container you will probably find a few bees lying dead, but this is to be expected after the stressful journey that they have undertaken. The rest of the bees will be alright and will have to be transferred to the new hive that you have set up for them from the container that they were shipped in. Before trying to transfer the bees make sure that you have the safety gear on and also the smoker ready. Inside the shipping container there will be a smaller container; this has the queen bee in it. This box will be closed with a cork, and if you remove the cork there will be another stopper inside that is made of sugar.

Placing the queen in the hive

The queen's container should be hung inside the hive that has been prepared for the new bees. Now pierce a hole in the sugar so that the worker bees will be able to free the queen bee easier and allow her to escape into the hive. Care should be taken that the queen bee is not damaged in anyway while you pierce the sugar cube as it is not easy to find a replacement for the queen bee during winter months.

After the queen bee has been put into the hive, blow a puff of smoke into the container with the other bees and allow them out into the hive. The bees will automatically spill out of the container into the hive and settle down there. Make sure you put a feeder filled with ordinary sugar n to the hive. If there are any bees still in the container just leave the container near the hive and they will go into it on their own. The bees prefer being changed from the container to the hive wither early in the morning or in the evening time.

A week for the bees to settle down

The bees will take at least a week to settle down to their new hive and then the queen bee will start laying her eggs and the bees will also start making honey.

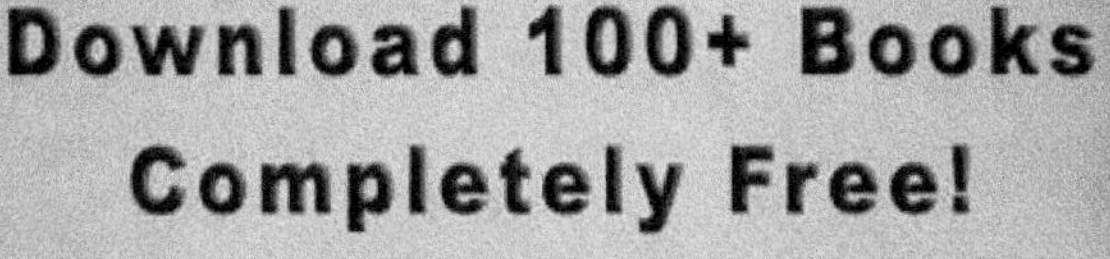

Download 100+ Books
Completely Free!
No Hassles...
No Charges...
Just Click And Download

100% Free!

CLICK HERE

This Product Is Brought To You By